Inside the Mind of a Psychopath: Exploring the Psychology and Behavior of Psychopaths

By NSKR.

_____

Table of Contents

Author's Explanation:

As the author of "Inside the Mind of a Psychopath: Exploring the Psychology and Behavior of Psychopaths," I embarked on this journey with a profound curiosity and a deep-seated commitment to unraveling the mysteries surrounding psychopathy. I have dedicated years of study and research to understanding the intricacies of human behavior, particularly as they manifest in individuals diagnosed with psychopathy.

The motivation behind this book stems from a desire to bridge the gap between academic research and public understanding, to demystify the complexities of psychopathy, and to offer insights that may aid in the prevention, detection, and management of this often misunderstood personality disorder.

Each chapter of this book is meticulously crafted to provide readers with a comprehensive overview of psychopathy, drawing from a rich tapestry of historical accounts, empirical research, clinical insights, and real-world examples. Through a multidisciplinary lens, I endeavor to explore the nuances of psychopathy, from its neurobiological underpinnings to its societal implications.

In delving into the psychology of psychopaths, my aim is not only to shed light on the inner workings of the psychopathic mind but also to foster empathy, understanding, and informed discourse surrounding this complex phenomenon. By challenging misconceptions and confronting stigma, I hope to pave the way for greater awareness and compassion in our interactions with individuals who may exhibit psychopathic traits.

As readers embark on this journey of exploration, I encourage them to approach the topic with an open mind and a critical eye, to question assumptions, and to engage with the material in a spirit of inquiry and reflection. It is my sincere hope that this book serves as a catalyst for deeper conversations, further research, and ultimately, positive change in how we perceive and address psychopathy in our society.

Thank you for joining me on this intellectual odyssey. May our collective efforts contribute to a more nuanced understanding of the human psyche and pave the way toward a future marked by empathy, insight, and resilience.

Warm regards,

NSKR

Chapter 1: Introduction to Psychopathy

Psychopathy is a term that often invokes images of cunning criminals and master manipulators, but its roots delve much deeper into the complexities of human behavior and psychology. In this chapter, we embark on a journey to understand the essence of psychopathy, tracing its origins and exploring its multifaceted nature.

The Origins of Psychopathy

The term "psychopathy" has its origins in the Greek words "psyche," meaning mind, and "pathos," meaning suffering or disease. Early conceptions of psychopathy date back to the 19th century, with pioneers such as Philippe Pinel and Benjamin Rush laying the groundwork for its study. Pinel described individuals who exhibited a lack of moral conscience and empathy, while Rush coined the term "moral insanity" to describe similar behavioral patterns.

Historical Perspectives

Throughout history, perceptions of psychopathy have evolved alongside advancements in psychology and psychiatry. From the moral degeneracy theories of the 19th century to the psychodynamic and behavioral perspectives of the 20th century, various schools of thought have sought to explain the origins and manifestations of psychopathic traits.

The Emergence of Modern Concepts

The modern conceptualization of psychopathy owes much to the pioneering work of researchers such as Hervey Cleckley and Robert Hare. Cleckley's seminal work, "The Mask of Sanity," provided a groundbreaking exploration of psychopathic personality traits, emphasizing superficial charm, egocentricity, and a lack of empathy as core features. Hare's Psychopathy Checklist-Revised (PCL-R) further refined the assessment and diagnosis of psychopathy, setting a standard for research and clinical practice.

Defining Psychopathy

Psychopathy is characterized by a distinctive set of personality traits and behavioral patterns that differentiate it from other mental health disorders. Central to the concept of psychopathy are traits such as grandiosity, manipulation, impulsivity, and a lack of remorse or empathy. Individuals with psychopathic traits often exhibit a charming and charismatic facade, masking their underlying callousness and disregard for others' well-being.

Case Study: The Case of John

Consider the case of John, a successful businessman with a charming demeanor and magnetic personality. On the surface, John appears to be a model citizen, admired by colleagues and respected in his community. However, beneath his polished exterior lies a darker truth. John exhibits a pattern of deceit, manipulation, and exploitation in his interpersonal relationships and business dealings. Despite his outward charm, he lacks empathy and remorse for the harm he causes others, viewing them merely as pawns in his pursuit of personal gain. John's behavior aligns closely with the traits and characteristics associated with psychopathy, illustrating the complexities of this enigmatic personality disorder.

Key Concepts and Takeaways

- Psychopathy is a multifaceted personality disorder characterized by traits such as superficial charm, manipulativeness, and a lack of empathy.

- Historical perspectives and modern research have contributed to our understanding of psychopathy and its manifestations.

- The emergence of standardized assessment tools, such as the PCL-R, has facilitated the diagnosis and study of psychopathy in clinical and research settings.

- Case studies, such as the case of John, provide concrete examples of psychopathic behavior and its impact on individuals and society.

In the chapters that follow, we will delve deeper into the complexities of psychopathy, exploring its neurobiological underpinnings, its implications for criminal behavior and social functioning, and the challenges and opportunities inherent in its diagnosis and treatment. Through our exploration, we aim to unravel the mysteries of the psychopathic mind and shed light on this compelling yet often misunderstood aspect of human psychology.

—

Chapter 2: The History of Psychopathy

Understanding the history of psychopathy provides crucial insights into its evolution as a concept within psychology and society. From ancient civilizations to modern research, the study of psychopathy has undergone significant transformations, shaped by cultural, philosophical, and scientific influences.

---

1. Ancient Roots

The roots of psychopathy trace back to ancient civilizations, where descriptions of antisocial behavior and moral deviance can be found in historical texts and religious scriptures. In ancient Greece, philosophers such as Plato and Aristotle pondered the nature of evil and the existence of individuals devoid of conscience or empathy.

Example: In Plato's "Republic," the character of Gyges illustrates the moral corruption that arises when individuals are granted power without accountability, a theme that resonates with contemporary understandings of psychopathy and its implications for society.

---

2. Early Medical and Scientific Perspectives

The emergence of modern medicine and scientific inquiry during the Renaissance laid the groundwork for early theories of mental illness and abnormal behavior. Influential figures like Philippe Pinel and

Benjamin Rush contributed to the classification and study of psychological disorders, including forms of psychopathy.

Example: Benjamin Rush, often regarded as the "Father of American Psychiatry," described psychopathic individuals as exhibiting a lack of moral sense and a propensity for deceitful and manipulative behavior, foreshadowing later diagnostic criteria for psychopathy.

---

3. The Rise of Psychiatric Institutions

The 19th century witnessed the establishment of psychiatric institutions and the professionalization of mental health care. Pioneers such as Emil Kraepelin and Cesare Lombroso explored the nature of criminal behavior and personality disorders, laying the foundation for the modern concept of psychopathy.

Example: Cesare Lombroso's theory of the "born criminal" proposed a biological basis for criminal behavior, suggesting that individuals with certain physical traits were predisposed to antisocial conduct. While his ideas have been largely discredited, they reflect early attempts to understand the etiology of psychopathy.

---

4. The Influence of Psychiatry and Psychology

The 20th century saw significant advancements in the fields of psychiatry and psychology, leading to the refinement of diagnostic criteria and assessment tools for psychopathy. Figures such as Hervey Cleckley and Robert Hare played pivotal roles in defining and studying psychopathic personality traits.

Example: Hervey Cleckley's seminal work, "The Mask of Sanity," introduced the concept of the psychopathic personality and described the superficial charm, lack of empathy, and pathological lying characteristic of psychopathic individuals. His clinical insights laid the groundwork for future research and diagnosis of psychopathy.

---

5. Psychopathy in Criminology and Forensic Psychology

Psychopathy became a central focus in criminology and forensic psychology, with researchers exploring its prevalence among criminal populations and its implications for recidivism and violent behavior.

Landmark studies such as the work of Robert Hare on the Psychopathy Checklist revolutionized the assessment and understanding of psychopathy in forensic settings.

Example: The case of Ted Bundy, a notorious serial killer and psychopath, captured the public's attention and highlighted the dangers posed by individuals with psychopathic traits. Bundy's ability to charm and manipulate his victims underscored the deceptive nature of psychopathy and its devastating consequences.

---

6. Contemporary Perspectives and Challenges

In the 21st century, research on psychopathy continues to evolve, with interdisciplinary approaches integrating neuroscience, genetics, and social psychology to unravel its complexities. However, controversies surrounding the diagnosis and treatment of psychopathy persist, raising ethical and practical dilemmas for clinicians and policymakers alike.

Example: The debate over psychopathy and its relationship to criminal responsibility and moral culpability remains contentious, with legal systems grappling with questions of punishment versus rehabilitation for psychopathic offenders. High-profile cases, such as those involving psychopathic individuals pleading insanity, highlight the ongoing challenges in addressing psychopathy within the criminal justice system.

Conclusion

The history of psychopathy reflects a rich tapestry of cultural, philosophical, and scientific influences that have shaped our understanding of this complex phenomenon. From ancient philosophy to modern neuroscience, the study of psychopathy continues to captivate scholars and practitioners, driving advancements in research, diagnosis, and treatment. By examining its historical roots and evolutionary trajectory, we gain valuable insights into the enduring mysteries of the human psyche and the challenges posed by individuals who defy conventional notions of morality and conscience.

___

Chapter 3: Understanding the Psychopathic Mind

The psychopathic mind is a complex landscape of cognitive and emotional processes that set individuals with this personality disorder apart from the general population. Understanding these intricacies is crucial for recognizing and navigating the behaviors and motivations of psychopathic individuals. In this chapter, we will delve deeper into the various facets of the psychopathic mind, drawing upon real-life examples and case studies to illuminate key concepts.

Section 1: The Mask of Charm

One of the defining characteristics of psychopathy is the ability to charm and manipulate others effortlessly. Psychopathic individuals often possess a superficial charm that masks their true intentions and motivations. Consider the case of Ted Bundy, one of the most notorious serial killers in history. Bundy's charming demeanor and charisma allowed him to lure numerous victims to their deaths, all while maintaining a façade of normalcy in his public life.

Case Study: Ted Bundy

Ted Bundy was a charming and intelligent young man who used his charisma to gain the trust of his victims. Despite his outward charm and affability, Bundy harbored dark impulses and a complete lack of empathy for his victims. His ability to charm and manipulate those around him enabled him to evade capture for years, leaving a trail of devastation in his wake.

Section 2: Lack of Empathy

A hallmark trait of psychopathy is a profound lack of empathy for others. Psychopathic individuals are unable to experience genuine empathy or compassion, viewing others as mere objects to be exploited for their own gain. This lack of empathy allows psychopaths to commit heinous acts without remorse or guilt, making them particularly dangerous individuals.

Case Study: Aileen Wuornos

Aileen Wuornos, a notorious serial killer, exemplifies the lack of empathy characteristic of psychopathy. Wuornos, a former prostitute, murdered seven men in cold blood, claiming that she acted in self-defense. Despite her claims, Wuornos showed no remorse for her actions and displayed a complete disregard for the lives she had taken. Her lack of empathy and remorse marked her as a classic psychopath.

Section 3: Impulsivity and Risk-Taking Behavior

Psychopathic individuals often exhibit a pattern of impulsivity and reckless behavior, driven by a desire for stimulation and excitement. These individuals may engage in high-risk activities without considering the potential consequences, leading to a cycle of thrill-seeking and impulsive decision-making.

Case Study: John Wayne Gacy

John Wayne Gacy, also known as the "Killer Clown," epitomized the impulsivity and risk-taking behavior associated with psychopathy. Gacy, a respected member of his community, harbored a dark secret: he was responsible for the sexual assault and murder of at least 33 young men and boys. Gacy's penchant for taking risks and engaging in dangerous behavior ultimately led to his downfall, as law enforcement uncovered the gruesome truth behind his outward façade.

Section 4: Manipulative Tactics

Psychopathic individuals are masters of manipulation, using cunning tactics to exploit and control those around them. Whether through lies, deceit, or manipulation, psychopaths are adept at getting what they want at the expense of others.

Case Study: Bernie Madoff

Bernie Madoff, the mastermind behind one of the largest Ponzi schemes in history, exemplifies the manipulative tactics employed by psychopathic individuals. Madoff, a highly respected financier, defrauded thousands of investors out of billions of dollars through a web of lies and deception. Despite his charming demeanor and reputation for financial acumen, Madoff's true nature as a psychopath was revealed when his elaborate scheme unraveled, leaving countless lives shattered in its wake.

Section 5: Lack of Remorse and Accountability

One of the most disturbing aspects of psychopathy is the complete absence of remorse or accountability for one's actions. Psychopathic individuals are unable to acknowledge the harm they have caused or take responsibility for their behavior, leading to a pattern of repeated offenses and a lack of meaningful change.

Case Study: Jeffrey Dahmer

Jeffrey Dahmer, infamously known as the "Milwaukee Cannibal," epitomizes the lack of remorse and accountability characteristic of psychopathy. Dahmer murdered and dismembered 17 young men and boys, displaying a complete disregard for human life. Despite his heinous crimes, Dahmer showed little remorse for his actions, even attempting to rationalize his behavior in interviews with law enforcement. His utter lack of remorse and accountability underscored his psychopathic nature and the depths of his depravity.

In conclusion, the psychopathic mind is a complex and enigmatic entity characterized by charm, manipulation, lack of empathy, impulsivity, and a profound disregard for others. By examining real-life examples and case studies, we gain valuable insights into the inner workings of psychopathy and the devastating impact it can have on individuals and society as a whole. Understanding the psychopathic mind is essential for recognizing the warning signs, protecting ourselves and others, and ultimately working towards prevention and intervention strategies to address this pervasive personality disorder.

Chapter 4: The Traits and Characteristics of Psychopaths

Understanding psychopathy begins with a deep dive into the traits and characteristics that define this complex personality disorder. Psychopaths exhibit a distinct set of behaviors and tendencies that set them apart from the general population. In this chapter, we explore these traits in detail, drawing from research, case studies, and real-life examples to illuminate the essence of psychopathy.

---

1. Superficial Charm

One of the hallmark traits of psychopathy is superficial charm. Psychopaths are often adept at charming others and presenting a charismatic facade, masking their true intentions and underlying lack of empathy. For example, a psychopathic individual may effortlessly charm their way into positions of authority or manipulate others to achieve their goals, all while concealing their true motives behind a veneer of charm and charisma.

Case Study: James, a successful businessman, was known for his charming personality and magnetic charisma. He effortlessly captivated colleagues and clients alike with his smooth-talking demeanor and confident demeanor. However, behind the charm lay a calculated manipulator who exploited others for personal gain, demonstrating a classic example of psychopathic charm in action.

---

2. Lack of Empathy

Perhaps the most defining trait of psychopathy is a profound lack of empathy. Psychopaths display a callous disregard for the feelings and well-being of others, viewing them as mere objects to be used and discarded for their own benefit. This absence of empathy allows psychopaths to engage in harmful behaviors without experiencing guilt or remorse.

Case Study: Sarah, a diagnosed psychopath, showed a complete disregard for the feelings of others from a young age. Despite witnessing the distress of her classmates when she bullied them, Sarah remained indifferent to their suffering, deriving pleasure from their pain. Her lack of empathy enabled her to inflict harm without experiencing any sense of remorse or guilt, highlighting a core feature of psychopathy.

---

3. Manipulative Behavior

Psychopaths are skilled manipulators who excel at exploiting the vulnerabilities of others for personal gain. They use charm, deception, and manipulation to control and manipulate their victims, often with devastating consequences. Whether in personal relationships, professional settings, or criminal enterprises, psychopaths employ manipulation as a powerful tool to achieve their objectives.

Case Study: Mark, a psychopathic con artist, masterfully manipulated his victims through a combination of charm and deception. Posing as a wealthy businessman, Mark lured unsuspecting investors into fraudulent schemes, promising lucrative returns on their investments. Through his persuasive tactics and false promises, Mark swindled millions of dollars from his victims, leaving behind a trail of devastation and financial ruin.

---

4. Impulsivity and Risk-taking

Psychopaths exhibit a tendency towards impulsivity and risk-taking behavior, often engaging in reckless and irresponsible actions without considering the consequences. Whether it's substance abuse, criminal activity, or thrill-seeking behavior, psychopaths are drawn to high-risk activities that provide stimulation and excitement.

Case Study: Emily, a diagnosed psychopath, was notorious for her reckless behavior and disregard for personal safety. Despite numerous warnings and interventions, Emily continued to engage in dangerous activities, such as reckless driving and substance abuse, without considering the potential consequences. Her impulsivity and thrill-seeking tendencies ultimately led to a series of negative outcomes, highlighting the inherent risks associated with psychopathic behavior.

---

5. Lack of Long-term Goals and Stability

Psychopaths often exhibit a lack of long-term goals and stability in their lives, leading to a pattern of erratic behavior and impulsivity. They may struggle to maintain stable relationships, employment, and financial stability due to their short-sighted focus on immediate gratification and self-interest.

Case Study: David, a diagnosed psychopath, struggled to maintain stability in his personal and professional life. Despite possessing intelligence and charm, David's impulsive nature and lack of long-term planning led to a series of failed relationships and job losses. His inability to prioritize long-term goals and commitments resulted in a cycle of instability and chaos, reflecting a common feature of psychopathic individuals.

--

Conclusion

The traits and characteristics of psychopaths paint a complex portrait of individuals who possess a unique blend of charm, manipulation, impulsivity, and callousness. By examining these traits through real-life examples and case studies, we gain valuable insights into the nature of psychopathy and its profound impact on individuals and society. Understanding these traits is essential for recognizing and addressing the challenges posed by psychopathy in various domains of life, from interpersonal relationships to criminal justice systems.

—

Chapter 5: Nature vs. Nurture: The Debate on Psychopathy

Understanding the origins of psychopathy is a complex endeavor that involves unraveling the interplay between genetic predispositions and environmental influences. The age-old debate of nature versus nurture looms large in the study of psychopathy, as researchers strive to untangle the contributions of biology and upbringing to the development of this personality disorder.

Section 1: The Role of Genetics in Psychopathy

Genetics play a significant role in shaping individual differences in personality and behavior, including the predisposition to psychopathy. Twin studies have provided compelling evidence for the heritability of psychopathic traits, suggesting that genetic factors contribute to the manifestation of psychopathy to a significant degree.

Example: In a landmark twin study conducted by researchers at the University of Minnesota, identical twins raised apart showed a higher concordance rate for psychopathic traits compared to fraternal twins, supporting the influence of genetic factors in the development of psychopathy.

Section 2: Environmental Influences on Psychopathy

While genetics lay the foundation for psychopathic tendencies, environmental factors also play a crucial role in shaping the expression and severity of psychopathy. Adverse childhood experiences, such as neglect, abuse, and trauma, have been consistently linked to the development of psychopathic traits later in life.

Case Study: The case of James, a young boy raised in a dysfunctional family environment characterized by parental neglect and substance abuse, exemplifies the profound impact of adverse childhood experiences on the development of psychopathy. Despite exhibiting early signs of emotional detachment and callousness, James received minimal support and intervention, eventually leading to escalating antisocial behavior and criminal involvement in adolescence.

Section 3: Gene-Environment Interactions

The interaction between genetic predispositions and environmental influences shapes the trajectory of psychopathy, with certain individuals being more vulnerable to adverse environmental factors based on their genetic makeup. Gene-environment interaction studies seek to unravel the intricate interplay between genetic vulnerabilities and environmental stressors in the development of psychopathy.

Example: Research conducted by Dr. Adrian Raine and colleagues at the University of Pennsylvania revealed that individuals with a combination of genetic risk factors, such as low monoamine oxidase (MAO) activity, and adverse environmental experiences, such as childhood maltreatment, were more likely to exhibit psychopathic traits in adulthood compared to those with either genetic or environmental risk factors alone.

Section 4: Epigenetics and Psychopathy

Epigenetic mechanisms, which involve modifications to gene expression without altering the underlying DNA sequence, provide further insights into the complex interplay between nature and nurture in psychopathy. Environmental influences can induce epigenetic changes that influence the expression of genes associated with psychopathic traits, potentially contributing to the development and maintenance of psychopathy.

Case Study: Sarah, a young woman with a family history of psychopathy, grew up in a stable and nurturing environment. Despite her genetic predisposition, Sarah's supportive upbringing and positive social influences facilitated adaptive coping mechanisms and emotional regulation skills, mitigating the expression of psychopathic traits and promoting prosocial behavior.

Section 5: Implications for Intervention and Prevention

The recognition of the dual influences of nature and nurture on psychopathy has profound implications for intervention and prevention efforts. Early identification of genetic vulnerabilities and environmental risk factors, coupled with targeted interventions aimed at addressing underlying mechanisms and promoting resilience, holds promise for reducing the prevalence and severity of psychopathy.

Example: The implementation of early intervention programs that target at-risk children and families, such as parent training, family therapy, and trauma-informed care, can mitigate the impact of adverse childhood experiences and foster healthy socioemotional development, thereby reducing the likelihood of psychopathy later in life.

Conclusion

The debate on the relative contributions of nature and nurture to psychopathy remains a central focus of research and debate in the field of psychology. While genetic predispositions lay the foundation for psychopathic tendencies, environmental influences play a critical role in shaping the expression and severity of psychopathy. By unraveling the complex interplay between genetic vulnerabilities and environmental stressors, researchers and practitioners strive to develop targeted interventions and prevention strategies that address the root causes of psychopathy and promote positive outcomes for individuals and society as a whole.

––––

Chapter 6: The Development of Psychopathy: From Childhood to Adulthood

Understanding the development of psychopathy is crucial for early intervention and prevention efforts. This chapter explores the intricate pathways that lead to the emergence of psychopathic traits, from childhood precursors to adult manifestations, through the lens of research, case studies, and real-life examples.

Section 1: Early Signs and Precursors

Psychopathy often begins to manifest in childhood, albeit in subtle and sometimes overlooked ways. Early signs and precursors, such as callousness, impulsivity, and conduct problems, may foreshadow the development of psychopathic traits later in life.

Example: A young child who exhibits a lack of remorse or empathy for others, engages in aggressive behavior without provocation, and displays manipulative tendencies may demonstrate early signs of psychopathy. These behaviors, if left unaddressed, can escalate over time and contribute to the development of full-blown psychopathy in adolescence or adulthood.

Section 2: The Role of Genetics and Environment

The development of psychopathy is influenced by a complex interplay of genetic and environmental factors. Genetic predispositions, coupled with adverse childhood experiences such as abuse, neglect, and unstable family environments, contribute to the formation of psychopathic traits and behaviors.

Example: Studies have shown that individuals with a family history of psychopathy or antisocial behavior are at higher risk of developing psychopathic traits themselves, suggesting a genetic component to the disorder. However, environmental factors such as exposure to violence, trauma, and dysfunctional family dynamics can also shape the development of psychopathy, highlighting the importance of early intervention and supportive interventions.

Section 3: Neurodevelopmental Processes

Neurobiological factors play a crucial role in the development of psychopathy. Dysfunctions in key brain regions implicated in emotional processing, impulse control, and moral reasoning contribute to the emergence of psychopathic traits and behaviors.

Example: Neuroimaging studies have identified abnormalities in the amygdala, prefrontal cortex, and other brain regions associated with empathy, emotion regulation, and moral decision-making in individuals with psychopathy. These neurodevelopmental disruptions may underlie deficits in empathy, remorse, and moral judgment observed in psychopathic individuals, shedding light on the biological underpinnings of the disorder.

Section 4: Developmental Trajectories

Psychopathy follows distinct developmental trajectories characterized by escalating antisocial behavior, interpersonal dysfunction, and disregard for societal norms. Understanding these trajectories can inform early intervention efforts and mitigate the risk of adverse outcomes.

Example: Longitudinal studies have identified different developmental pathways leading to psychopathy, including the childhood-onset and adolescent-onset trajectories. Children who exhibit early-onset conduct problems, aggression, and callous-unemotional traits are more likely to follow a persistent and severe trajectory of psychopathy into adulthood, whereas those with adolescent-onset behavior may display milder psychopathic features that emerge later in adolescence or early adulthood.

Section 5: Protective Factors and Resilience

While the development of psychopathy is influenced by a myriad of risk factors, protective factors and resilience can mitigate its impact and promote positive outcomes. Supportive relationships, positive role models, and access to resources and interventions play a crucial role in buffering against the adverse effects of psychosocial stressors.

Example: Research has shown that children who experience stable and nurturing relationships with caregivers, receive early intervention services, and participate in prosocial activities are less likely to develop psychopathic traits and engage in antisocial behavior later in life. These protective factors foster resilience and adaptive coping mechanisms, mitigating the risk of psychopathy and promoting healthy development.

Section 6: Case Studies and Clinical Illustrations

Examining real-life case studies and clinical illustrations provides valuable insights into the complex and multifaceted nature of psychopathy. Through in-depth analysis and exploration of individual experiences, we gain a deeper understanding of the developmental pathways, risk factors, and challenges associated with psychopathy.

Case Study 1: John, a 10-year-old boy, exhibits a pattern of aggressive behavior, deceitfulness, and disregard for rules and authority figures at home and school. Despite interventions and disciplinary measures, his behavior continues to escalate, leading to conflicts with peers and authority figures. Further assessment reveals underlying callous-unemotional traits and deficits in empathy and remorse, suggesting early signs of psychopathy. Early intervention and targeted therapies focusing on emotion regulation and social skills development are recommended to address John's behavioral challenges and mitigate the risk of psychopathy in adolescence and adulthood.

Case Study 2: Sarah, a teenager with a history of trauma and familial instability, displays impulsive behavior, substance abuse, and involvement in delinquent activities. Despite efforts to provide support and guidance, Sarah's behavior becomes increasingly reckless and antisocial, resulting in legal consequences and strained relationships with family and peers. A comprehensive assessment reveals underlying psychopathic traits, including manipulativeness, lack of empathy, and shallow affect. Intensive intervention and therapeutic approaches focusing on trauma-informed care, cognitive-behavioral therapy, and family-based interventions are recommended to address Sarah's complex needs and promote positive outcomes.

Conclusion

The development of psychopathy is a complex and multifaceted process shaped by genetic, neurobiological, and environmental factors. By understanding the early signs, developmental trajectories, and risk and protective factors associated with psychopathy, we can inform early intervention efforts, promote resilience, and mitigate the adverse effects of the disorder on individuals and society. Through continued research, awareness, and targeted interventions, we strive to foster healthy development and prevent the emergence of psychopathy in vulnerable populations.

—

Chapter 7: Psychopathy and Neurobiology: Unraveling the Mysteries of the Psychopathic Brain

---

1. Introduction to Neurobiology and Psychopathy

The study of psychopathy has increasingly turned towards understanding its neurobiological underpinnings. This chapter explores the intricate relationship between brain functioning and psychopathic traits, shedding light on the neural mechanisms that contribute to the characteristic behaviors and deficits observed in psychopathic individuals.

---

2. The Brain Regions Implicated in Psychopathy

Research using neuroimaging techniques such as functional magnetic resonance imaging (fMRI) and positron emission tomography (PET) has identified specific brain regions associated with psychopathy. The prefrontal cortex, amygdala, and anterior cingulate cortex are among the key areas implicated in emotional processing, decision-making, and empathy—functions that are often impaired in psychopathic individuals.

Example: Studies have shown reduced activity in the prefrontal cortex of psychopathic individuals during tasks involving moral reasoning and response inhibition, suggesting deficits in impulse control and decision-making processes.

---

3. Neurotransmitters and Neurochemistry in Psychopathy

Neurotransmitters, the chemical messengers of the brain, play a crucial role in regulating mood, behavior, and cognition. Dysregulation of neurotransmitter systems, including serotonin, dopamine, and norepinephrine, has been implicated in psychopathy. Imbalances in these neurotransmitters may contribute to the emotional dysregulation and reward-seeking behaviors characteristic of psychopathic individuals.

Example: Research has shown that psychopathic individuals exhibit reduced levels of serotonin, which is associated with impulsivity, aggression, and impaired emotional processing.

---

4. Structural and Functional Connectivity Differences

Beyond individual brain regions, studies have investigated the structural and functional connectivity patterns in the brains of psychopathic individuals. Differences in white matter integrity and connectivity between brain regions may underlie deficits in information processing, emotional regulation, and social cognition observed in psychopathy.

Example: Diffusion tensor imaging (DTI) studies have revealed alterations in the structural connectivity of the amygdala and prefrontal cortex in psychopathic individuals, suggesting disrupted communication between regions involved in emotion processing and cognitive control.

---

5. Genetic Influences on Neurobiology and Psychopathy

Genetic factors also play a role in shaping the neurobiology of psychopathy. Twin and family studies have demonstrated a significant heritable component to psychopathic traits, with variations in genes related to neurotransmitter function, brain development, and emotional regulation contributing to individual differences in susceptibility to psychopathy.

Example: Variants of the MAOA gene, which regulates the metabolism of neurotransmitters such as serotonin, have been linked to increased risk of psychopathy and antisocial behavior, particularly in individuals exposed to adverse environmental conditions.

---

6. Developmental Trajectories and Neurobiological Risk Factors

The interplay between genetic predispositions and environmental influences shapes the developmental trajectories of psychopathy. Early experiences of trauma, neglect, and adversity can interact with genetic vulnerabilities to alter neurobiological processes and increase the risk of developing psychopathic traits later in life.

Example: Longitudinal studies have shown that children exposed to chronic stress or maltreatment exhibit alterations in brain structure and function, including changes in the amygdala and prefrontal cortex, which may contribute to the development of psychopathic traits.

---

7. Intervention and Treatment Implications

Understanding the neurobiological basis of psychopathy has important implications for intervention and treatment strategies. Targeted interventions aimed at addressing neurobiological deficits, regulating emotional responses, and promoting prosocial behavior may help mitigate the harmful effects of psychopathy and reduce the risk of recidivism among affected individuals.

Example: Neurofeedback techniques, which use real-time monitoring of brain activity to train individuals to regulate their neural responses, show promise in improving emotional regulation and impulse control in psychopathic individuals.

---

8. Ethical Considerations and Future Directions

As research in the neurobiology of psychopathy progresses, ethical considerations regarding the use of neuroscientific evidence in legal and forensic contexts become increasingly relevant. It is essential to balance scientific advancements with ethical principles and considerations of individual rights and dignity.

Example: The use of neurobiological evidence in courtrooms to inform sentencing and parole decisions raises questions about the reliability, validity, and interpretation of such evidence, as well as its potential impact on perceptions of culpability and rehabilitation.

Conclusion

The study of psychopathy from a neurobiological perspective offers valuable insights into the underlying mechanisms and vulnerabilities associated with this complex personality disorder. By elucidating the neural correlates of psychopathy, researchers aim to inform more targeted and effective interventions, enhance our understanding of the disorder, and ultimately improve outcomes for individuals affected by psychopathy and society as a whole.

___

Chapter 8: Psychopathy in Popular Culture

Introduction to Psychopathy in Popular Culture

Psychopathy has long captured the imagination of artists, writers, filmmakers, and audiences around the world. From literary villains to cinematic antiheroes, psychopathic characters often serve as captivating figures that both fascinate and repel. In this chapter, we explore the portrayal of psychopathy in popular culture, examining the myths, misconceptions, and realities that shape our perceptions of this complex personality disorder.

Section 1: The Literary Landscape of Psychopathy

Literature has provided a fertile ground for the exploration of psychopathy, with countless novels featuring characters who exhibit psychopathic traits. From classic works of literature to contemporary bestsellers, authors have crafted compelling narratives that delve into the darkest recesses of the human psyche.

Example 1: "American Psycho" by Bret Easton Ellis

In Bret Easton Ellis's iconic novel "American Psycho," we are introduced to Patrick Bateman, a wealthy and successful investment banker who leads a double life as a sadistic serial killer. Bateman's outward charm and charisma mask his inner emptiness and propensity for violence, offering a chilling portrayal of psychopathy in the world of corporate excess.

Example 2: "Gone Girl" by Gillian Flynn

In Gillian Flynn's bestselling thriller "Gone Girl," the character of Amy Dunne emerges as a master manipulator and calculating sociopath. As the narrative unfolds, we are drawn into Amy's web of deception and manipulation, challenging our perceptions of trust, identity, and morality.

Section 2: Psychopathy on the Silver Screen

Film has also played a significant role in shaping our understanding of psychopathy, with memorable performances bringing psychopathic characters to life on the silver screen. From classic cinema to modern blockbusters, filmmakers have explored the darker aspects of human nature through compelling storytelling and visceral imagery.

Example 1: "The Silence of the Lambs" (1991)

In Jonathan Demme's psychological thriller "The Silence of the Lambs," Anthony Hopkins delivers a mesmerizing performance as Dr. Hannibal Lecter, a brilliant psychiatrist and cannibalistic serial killer. Lecter's cold intelligence and charismatic demeanor make him a captivating—and terrifying—figure, challenging the audience's perceptions of good and evil.

Example 2: "Joker" (2019)

Todd Phillips' "Joker" offers a gritty and unsettling portrayal of Arthur Fleck, a failed comedian who descends into madness and violence as he embraces his alter ego, the Joker. Joaquin Phoenix's haunting performance illuminates the character's fractured psyche, blurring the lines between victim and villain in a society plagued by corruption and inequality.

Section 3: Television and the Psychopathic Persona

Television series have also explored the intricacies of psychopathy, with complex characters and gripping storylines that keep audiences on the edge of their seats. From crime dramas to psychological thrillers, television offers a diverse array of narratives that delve into the minds of psychopathic individuals.

Example 1: "Dexter" (2006-2013)

In the television series "Dexter," Michael C. Hall portrays Dexter Morgan, a forensic blood spatter analyst for the Miami Metro Police Department who moonlights as a vigilante serial killer. Dexter's code of ethics, which dictates that he only targets other criminals, blurs the lines between justice and vigilantism, challenging viewers to confront their own moral compasses.

Example 2: "Mindhunter" (2017-2019)

"Mindhunter," created by Joe Penhall and based on the true-crime book "Mindhunter: Inside the FBI's Elite Serial Crime Unit" by John E. Douglas and Mark Olshaker, offers a chilling glimpse into the minds of serial killers through the eyes of FBI agents Holden Ford and Bill Tench. As they interview incarcerated murderers to understand their motives and methods, they confront the chilling reality of evil lurking within human nature.

Conclusion: Beyond the Screen

Psychopathy in popular culture serves as a mirror reflecting society's fascination with the darker aspects of human nature. While fictional portrayals often exaggerate and sensationalize psychopathic behavior for dramatic effect, they also raise important questions about morality, empathy, and the nature of evil. By exploring psychopathy through the lens of literature, film, and television, we gain valuable insights into the complexities of the human psyche and the enduring allure of the psychopathic persona.

—

Chapter 9: The Dark Triad: Psychopathy, Narcissism, and Machiavellianism

In this chapter, we explore the intricate dynamics of the Dark Triad—a cluster of personality traits characterized by psychopathy, narcissism, and Machiavellianism. Through in-depth analysis, case studies, and real-life examples, we delve into the manifestations, interconnections, and implications of these traits in various contexts.

Section 1: Understanding the Dark Triad

The Dark Triad represents a constellation of maladaptive personality traits that share commonalities in their core features:

1. **Psychopathy**: Characterized by a lack of empathy, impulsivity, and manipulative tendencies.

2. **Narcissism**: Defined by grandiosity, entitlement, and a need for admiration.

3. **Machiavellianism**: Marked by strategic manipulation, deceit, and a cynical worldview.

While each trait exhibits distinct characteristics, they often coalesce to form a potent combination of manipulative and exploitative behaviors.

Section 2: The Interplay Between Dark Triad Traits

The Dark Triad traits are not isolated entities but interconnected facets of personality that influence and reinforce each other's expression:

- **Example**: Consider a corporate executive who exhibits traits of psychopathy, narcissism, and Machiavellianism. His lack of empathy (psychopathy) enables him to exploit others for personal gain, while his grandiosity (narcissism) fuels his sense of entitlement and superiority. Additionally, his strategic manipulation (Machiavellianism) allows him to navigate power dynamics and achieve his objectives with little regard for ethical considerations.

Section 3: Dark Triad in Relationships and Interpersonal Dynamics

The presence of Dark Triad traits can significantly impact relationships and interpersonal dynamics, leading to manipulation, exploitation, and emotional harm:

- **Case Study**: Sarah, a young professional, enters into a romantic relationship with Mark, who exhibits traits of the Dark Triad. Initially charming and charismatic, Mark gradually reveals his true nature—manipulative, self-centered, and emotionally abusive. Despite Sarah's efforts to salvage the relationship, she becomes entangled in a web of deceit and manipulation, ultimately ending the toxic dynamic for her own well-being.

Section 4: Dark Triad in the Workplace

In corporate settings, individuals with Dark Triad traits may exploit organizational structures and interpersonal relationships for personal gain:

- **Example**: John, a high-ranking executive, employs Machiavellian tactics to undermine his colleagues and secure promotions. His narcissistic tendencies fuel his desire for recognition and power, while his lack of empathy allows him to disregard the well-being of others in pursuit of his goals. Despite his charming facade, John's manipulative behavior creates a toxic work environment characterized by distrust and discord.

Section 5: Recognizing and Addressing Dark Triad Traits

Identifying Dark Triad traits early is essential for mitigating their negative impact on individuals and organizations:

- **Case Study**: In a corporate environment, managers undergo training to recognize and address Dark Triad behaviors among employees. Through workshops and intervention programs, employees learn to identify manipulative tactics, set boundaries, and foster a culture of accountability and transparency. By promoting awareness and accountability, organizations can mitigate the harmful effects of Dark Triad traits and cultivate a healthier work environment.

Section 6: The Intersection of Dark Triad Traits and Psychopathology

While Dark Triad traits share similarities with psychopathy, they differ in their underlying motivations and expressions:

- **Example**: Psychopathy is characterized by a lack of remorse and empathy, whereas narcissism is driven by a sense of grandiosity and entitlement. Machiavellianism, on the other hand, is marked by strategic manipulation and deceit. Despite these distinctions, individuals with Dark Triad traits may exhibit overlapping patterns of behavior that pose challenges for diagnosis and treatment.

Conclusion

The Dark Triad represents a complex interplay of personality traits that exert a profound influence on individuals and society. By understanding the manifestations, interconnections, and implications of psychopathy, narcissism, and Machiavellianism, we can better navigate interpersonal dynamics, recognize warning signs, and foster environments that promote empathy, authenticity, and ethical conduct. Through continued research, awareness, and intervention, we strive to mitigate the negative impact of Dark Triad traits and promote psychological well-being in individuals and communities alike.

—

Chapter 10: The Relationship Between Psychopathy and Crime

Crime and psychopathy often intersect, presenting a complex and challenging landscape for researchers, law enforcement officials, and society as a whole. In this chapter, we delve deep into the intricate relationship between psychopathy and criminal behavior, exploring the underlying motivations, risk factors, and patterns of offending that characterize this dynamic connection.

Section 1: Understanding Psychopathic Offenders

Psychopathic individuals exhibit a distinct set of traits and characteristics that contribute to their involvement in criminal activities. From a lack of empathy and remorse to impulsivity and sensation-seeking behavior, these traits predispose psychopaths to engage in antisocial and often unlawful behavior.

Example: Consider the case of John, a diagnosed psychopath with a history of impulsive aggression and disregard for social norms. Despite knowing the consequences, John repeatedly engages in violent acts, demonstrating a callous disregard for the rights and well-being of others.

Section 2: Exploring the Motivations Behind Psychopathic Crime

The motivations driving psychopathic offenders to commit crimes are multifaceted and often rooted in a combination of personal gratification, power dynamics, and manipulation. For some psychopaths, criminal behavior serves as a means to assert dominance, achieve financial gain, or satisfy their need for excitement and stimulation.

Example: Sarah, a high-functioning psychopath, manipulates and deceives vulnerable individuals for financial gain. Through charm and persuasion, she exploits their trust and naivety, leaving a trail of victims in her wake.

Section 3: Risk Factors and Vulnerabilities

Psychopathy does not develop in isolation but is influenced by a myriad of genetic, environmental, and psychosocial factors. Early childhood trauma, adverse family dynamics, and genetic predispositions contribute to the emergence and exacerbation of psychopathic traits, increasing the likelihood of involvement in criminal behavior later in life.

Example: Michael, raised in an abusive and neglectful environment, exhibits early signs of callousness and aggression. As he grows older, his lack of empathy and disregard for authority lead him down a path of criminality, culminating in a series of violent offenses.

Section 4: Patterns of Offending

Psychopathic offenders display distinct patterns of offending characterized by impulsivity, calculated risk-taking, and a lack of remorse or empathy for their victims. From white-collar fraud to violent crimes, psychopaths are adept at adapting their tactics to exploit opportunities and evade detection.

Example: David, a white-collar criminal with psychopathic tendencies, embezzles millions of dollars from unsuspecting investors through elaborate Ponzi schemes. Despite the devastating consequences for his victims, David remains remorseless and unrepentant, prioritizing his own interests above all else.

Section 5: Challenges in Detection and Prosecution

Identifying and prosecuting psychopathic offenders present unique challenges for law enforcement agencies and the criminal justice system. The superficial charm and manipulative tactics employed by psychopaths often obscure their true intentions, making it difficult to establish guilt beyond a reasonable doubt.

Example: Lisa, a charismatic and well-respected community leader, conceals her psychopathic traits behind a façade of charm and likability. Despite mounting evidence of her involvement in a Ponzi scheme, Lisa's ability to manipulate and influence others complicates efforts to hold her accountable for her crimes.

Section 6: Intervention and Rehabilitation

Can psychopathic offenders be rehabilitated, or are they destined to repeat their criminal behavior indefinitely? The effectiveness of intervention programs and treatment approaches for psychopathy remains a topic of debate, with some experts advocating for early intervention and others emphasizing the limitations of current therapeutic modalities.

Example: Jake, a young offender diagnosed with psychopathy, participates in a specialized treatment program aimed at addressing his antisocial behavior and promoting prosocial skills. While initial progress is promising, Jake's resistance to change and persistent risk factors pose significant challenges to long-term rehabilitation efforts.

Section 7: Preventive Strategies and Public Policy

Preventing psychopathic individuals from engaging in criminal behavior requires a multifaceted approach that addresses underlying risk factors, enhances community support systems, and promotes early intervention and treatment. Public policies aimed at reducing recidivism and supporting reintegration into society play a crucial role in mitigating the impact of psychopathy on individuals and communities.

Example: The implementation of targeted prevention programs in high-risk communities helps identify and support individuals at risk of developing psychopathic traits. By providing access to mental health resources, educational opportunities, and social support networks, these initiatives empower individuals to break the cycle of criminality and pursue positive pathways.

Conclusion

The relationship between psychopathy and crime is a complex and multifaceted phenomenon that defies simple explanations. By understanding the motivations, risk factors, and patterns of offending associated with psychopathic individuals, we can develop more effective strategies for prevention, intervention, and rehabilitation. Through continued research, collaboration, and advocacy, we strive to create safer and more resilient communities that are equipped to address the challenges posed by psychopathy in all its forms.

—

Chapter 11: Psychopathy in Different Contexts: Corporate, Political, and Criminal

Psychopathy, with its distinct set of traits and characteristics, transcends individual behavior and permeates various aspects of society. In this chapter, we delve deeper into the manifestation of psychopathic traits in different contexts, namely corporate, political, and criminal settings. Through examples and case studies, we explore the impact of psychopathy on organizational dynamics, political decision-making, and criminal behavior.

Section 1: Psychopathy in the Corporate World

The corporate landscape is often characterized by competitiveness, ambition, and the pursuit of success. In such environments, psychopathic traits can flourish, manifesting in behaviors that prioritize self-interest over ethical considerations.

Case Study: The Enron Scandal

One of the most infamous examples of psychopathy in the corporate world is the Enron scandal. Enron, once hailed as a paragon of corporate success, collapsed in 2001 due to widespread accounting fraud and unethical business practices. Key figures within the organization, including CEO Jeffrey Skilling and CFO Andrew Fastow, exhibited psychopathic traits such as deceitfulness, manipulation, and a lack of empathy for employees and shareholders. The Enron case serves as a stark reminder of the destructive potential of psychopathy within corporate cultures.

Section 2: Psychopathy in Politics

Politics, with its power dynamics and high-stakes decision-making, provides fertile ground for the expression of psychopathic traits. Politicians who exhibit these traits may prioritize personal gain and political expedience over the welfare of their constituents and the common good.

Case Study: Richard Nixon

Richard Nixon, the 37th President of the United States, is often cited as a prime example of a political leader with psychopathic tendencies. Nixon's presidency was marked by a series of scandals, including the Watergate scandal, which ultimately led to his resignation in 1974. Nixon's actions, characterized by deceit, manipulation, and a disregard for democratic principles, underscore the potential consequences of psychopathy in positions of political authority.

Section 3: Psychopathy in Criminal Behavior

While not all individuals with psychopathic traits engage in criminal behavior, research suggests a significant overlap between psychopathy and antisocial conduct. From white-collar crime to violent offenses, psychopathy manifests in a variety of criminal contexts, posing challenges for law enforcement and criminal justice systems.

Case Study: Ted Bundy

Ted Bundy, one of the most notorious serial killers in modern history, epitomizes the intersection of psychopathy and criminal behavior. Bundy's charming facade and manipulative tactics allowed him to evade suspicion as he carried out a series of brutal murders across multiple states. His lack of remorse and empathy for his victims, coupled with a penchant for deception and manipulation, exemplifies the chilling nature of psychopathy in the realm of criminality.

Conclusion

The examples and case studies presented in this chapter offer a glimpse into the multifaceted nature of psychopathy and its impact across different domains of society. Whether in corporate boardrooms, political arenas, or criminal investigations, psychopathic traits can influence decision-making, interpersonal relationships, and societal dynamics in profound ways. By understanding the manifestations of psychopathy in various contexts, we gain valuable insights into the challenges posed by this complex personality disorder and the need for vigilance, awareness, and intervention in addressing its consequences.

—

Chapter 12: The Diagnosis and Assessment of Psychopathy

Diagnosing and assessing psychopathy is a complex and multifaceted process that requires a comprehensive understanding of the disorder's clinical features, behavioral manifestations, and underlying psychological mechanisms. In this chapter, we explore the various methods and tools used by clinicians to identify and evaluate psychopathy, drawing upon case studies and examples to illustrate key concepts and challenges.

Section 1: Understanding Psychopathy

Before delving into the diagnostic process, it is essential to grasp the fundamental characteristics and traits associated with psychopathy. Psychopathy is characterized by a constellation of features, including interpersonal manipulation, callousness, lack of empathy, impulsivity, and shallow affect. Individuals with psychopathic traits often exhibit a profound disregard for societal norms and the rights of others, leading to a pattern of persistent antisocial behavior.

Case Study 1: The Case of John

John is a 35-year-old man with a history of legal troubles, including multiple arrests for theft and assault. Despite repeated interventions and treatment attempts, he continues to engage in impulsive and reckless behavior, showing little remorse for the harm he causes to others. John's interpersonal relationships are characterized by manipulation and deceit, as he exploits others for personal gain without regard for their well-being.

Section 2: Assessment Tools and Measures

Assessing psychopathy requires the use of specialized tools and measures designed to capture the unique features of the disorder. One of the most widely used instruments for assessing psychopathy is the Psychopathy Checklist-Revised (PCL-R), developed by Robert Hare. The PCL-R consists of 20 items scored based on information obtained from interviews, behavioral observations, and collateral reports.

Case Study 2: Sarah's Assessment

Sarah, a forensic psychologist, conducts a comprehensive assessment of a client named Michael using the PCL-R. Through structured interviews and review of collateral information, Sarah gathers data on Michael's interpersonal style, emotional detachment, and antisocial behavior. She scores each item on the PCL-R based on the presence or absence of specific traits and behaviors, ultimately generating a total score that reflects Michael's level of psychopathy.

Section 3: Challenges and Controversies

Despite its widespread use, the assessment of psychopathy is not without challenges and controversies. Critics argue that psychopathy is a complex and heterogeneous construct that cannot be captured fully by a single measure or diagnostic tool. Moreover, concerns have been raised about the potential misuse of psychopathy assessments in forensic settings, where the stakes are high, and decisions about treatment, sentencing, and risk management are paramount.

Case Study 3: The Legal Implications

In a high-profile court case, the defense team argues that the use of psychopathy assessments, such as the PCL-R, is biased and unreliable, potentially leading to unjust outcomes for individuals with mental health disorders. The prosecution, however, maintains that psychopathy assessments provide valuable insights into an individual's risk of reoffending and are essential for informing sentencing and risk management decisions.

Section 4: Integrating Multiple Sources of Information

Effective assessment of psychopathy requires the integration of multiple sources of information, including clinical interviews, psychological testing, collateral reports, and behavioral observations. By triangulating data from diverse sources, clinicians can gain a more comprehensive understanding of an individual's psychopathic traits and their impact on functioning and behavior.

Case Study 4: The Case Conference

A team of forensic professionals convenes for a case conference to discuss the assessment findings for a client named Emily. Each team member presents their observations and impressions based on their respective evaluations, including interviews, psychological testing, and behavioral assessments. Through collaborative discussion and analysis, the team identifies key patterns and themes indicative of Emily's psychopathic traits, informing recommendations for treatment and risk management.

Section 5: Ethical Considerations and Professional Responsibility

In the assessment and diagnosis of psychopathy, clinicians must adhere to ethical guidelines and principles of professional conduct. This includes maintaining objectivity, safeguarding confidentiality, and ensuring the welfare and rights of clients. Additionally, clinicians must be mindful of the potential stigma and discrimination associated with psychopathy and take steps to mitigate harm and promote understanding and empathy.

Case Study 5: The Therapist's Dilemma

A therapist grapples with ethical dilemmas as she assesses a client named David for psychopathic traits. While David presents with several features consistent with psychopathy, the therapist recognizes the importance of approaching the assessment with empathy and compassion, refraining from stigmatizing or pathologizing his behavior. Through a client-centered and strengths-based approach, the therapist works collaboratively with David to address his underlying needs and promote positive change.

Conclusion

The diagnosis and assessment of psychopathy represent a critical step in understanding and addressing the complexities of this challenging disorder. By employing a multidimensional approach that integrates clinical expertise, empirical research, and ethical considerations, clinicians can provide comprehensive evaluations that inform treatment, intervention, and risk management strategies. Despite the inherent complexities and controversies, the assessment of psychopathy remains an essential endeavor in promoting the well-being and safety of individuals and communities alike.

—

Chapter 13: Treatment and Management of Psychopathy

Psychopathy poses unique challenges for treatment and management due to its complex nature and resistance to traditional therapeutic approaches. In this chapter, we explore the various strategies and interventions aimed at addressing psychopathic traits and reducing the risk of harmful behavior.

Section 1: Understanding Psychopathy Treatment

Effective treatment of psychopathy requires a multifaceted approach that addresses the underlying factors contributing to the disorder. Unlike other mental health conditions, psychopathy is characterized by a lack of empathy, shallow emotions, and a propensity for manipulative and antisocial behavior. Traditional psychotherapeutic methods may not be effective in treating psychopathy, as individuals with this disorder often lack insight into their own behavior and resist change.

Section 2: Treatment Approaches

1. **Cognitive-Behavioral Therapy (CBT)**: CBT aims to challenge and modify dysfunctional thought patterns and behaviors. While CBT has shown some effectiveness in reducing specific antisocial behaviors associated with psychopathy, its impact on core psychopathic traits such as lack of empathy and remorse is limited.

   Example: A case study involving a psychopathic individual participating in CBT may focus on identifying and challenging distorted thinking patterns related to entitlement and manipulation in interpersonal relationships. While the individual may show temporary improvement in specific behaviors, underlying psychopathic traits may remain largely unchanged.

2. **Dialectical Behavior Therapy (DBT)**: DBT combines elements of CBT with mindfulness techniques to help individuals regulate emotions

and improve interpersonal relationships. While DBT can be beneficial for managing impulsivity and emotional dysregulation commonly associated with psychopathy, its effectiveness in addressing core psychopathic traits remains uncertain.

Example: A psychopathic individual participating in a DBT program may learn mindfulness techniques to manage intense emotions and impulsive urges. However, the challenge lies in fostering genuine empathy and remorse, which are fundamental deficits in psychopathy.

3. **Risk Management and Supervision**: For individuals with psychopathy who pose a risk to themselves or others, risk management and supervision strategies may be necessary to prevent harm and maintain public safety. This may involve close monitoring, restrictions on access to potential victims, and intensive supervision within community or institutional settings.

Example: A psychopathic individual with a history of violent behavior may be placed under strict supervision and monitoring by mental health professionals and law enforcement agencies to mitigate the risk of future harm. This may include regular check-ins, electronic monitoring, and participation in structured treatment programs.

Section 3: Challenges and Limitations

1. **Resistance to Treatment**: Psychopathic individuals often exhibit resistance to treatment due to their egocentricity, lack of insight, and unwillingness to conform to societal norms. Attempts to impose change or instill empathy through traditional therapeutic methods may be met with defiance and manipulation.

   Example: Despite participating in therapy sessions, a psychopathic individual may feign compliance and manipulate the therapist to avoid confronting uncomfortable emotions or accepting responsibility for harmful actions. This resistance to change poses a significant barrier to effective treatment.

2. **High Rates of Recidivism**: Psychopathy is associated with an increased risk of recidivism and persistent antisocial behavior, even following interventions aimed at rehabilitation. The core features of psychopathy, including callousness and lack of empathy, contribute to ongoing patterns of misconduct and reoffending.

   Example: A psychopathic offender released from prison may exhibit temporary compliance with parole conditions but eventually revert to familiar patterns of deceit and manipulation to achieve personal gain.

Despite efforts to address specific risk factors, the underlying personality traits associated with psychopathy remain unchanged.

3. **Ethical Considerations**: The treatment and management of psychopathy raise ethical dilemmas related to autonomy, consent, and the potential for harm to self and others. Balancing the rights of individuals with psychopathy against the need to protect society from potential harm requires careful consideration of ethical principles and legal frameworks.

   Example: In cases where a psychopathic individual refuses treatment or poses a significant risk to public safety, ethical considerations may arise regarding involuntary commitment, coercive interventions, and the preservation of individual rights and dignity.

Section 4: Future Directions

Despite the challenges inherent in treating psychopathy, ongoing research and innovation offer hope for improved interventions and outcomes. Future directions in the treatment and management of psychopathy may include:

1. **Targeted Interventions**: Tailoring treatment approaches to address specific deficits and vulnerabilities associated with psychopathy, such as emotional processing deficits and interpersonal dysfunction.

2. **Early Intervention and Prevention**: Identifying at-risk individuals and implementing early intervention strategies to address psychopathic traits before they escalate into harmful behavior.

3. **Integration of Neurobiological Approaches**: Incorporating insights from neurobiology and neuroscience to develop targeted interventions that address underlying brain mechanisms implicated in psychopathy.

4. **Multidisciplinary Collaboration**: Fostering collaboration between mental health professionals, researchers, law enforcement agencies, and community stakeholders to develop comprehensive strategies for the treatment and management of psychopathy.

Conclusion

The treatment and management of psychopathy represent a complex and challenging endeavor that requires a nuanced understanding of the disorder and its underlying mechanisms. While traditional therapeutic approaches may offer limited efficacy in addressing core psychopathic traits, innovative strategies and multidisciplinary collaboration hold promise for improving outcomes and mitigating the impact of psychopathy on individuals and society. By recognizing the unique needs and challenges associated with psychopathy, we can strive to develop more effective interventions and support systems for individuals affected by this enigmatic personality disorder.

—

Chapter 14: Living with a Psychopath: Understanding and Coping Strategies

Living with or being in a relationship with a psychopath can be an emotionally and psychologically challenging experience. In this chapter, we will delve deeper into the dynamics of relationships involving psychopathic individuals, explore the impact on their partners, family members, and friends, and provide strategies for understanding and coping with the complexities of such relationships.

Understanding Psychopathic Relationships

Psychopathic individuals often exhibit superficial charm, manipulation, and a lack of empathy, making them skilled at drawing others into their orbit. Initially, relationships with psychopaths may appear exciting and intense, but over time, the true nature of their personality becomes apparent.

Case Study 1: Sarah and Mark

Sarah met Mark at a social gathering and was immediately drawn to his charismatic personality. He showered her with attention and affection, making her feel like the center of his world. However, as their relationship progressed, Sarah began to notice subtle signs of manipulation and deceit. Mark would frequently lie about his whereabouts and downplay Sarah's concerns, leaving her feeling confused and insecure.

The Impact of Psychopathy on Relationships

Living with a psychopath can take a toll on one's emotional well-being and mental health. The constant manipulation, gaslighting, and emotional abuse can erode self-esteem and create a sense of powerlessness in the relationship.

Case Study 2: John and Emily

Emily was initially captivated by John's confidence and charm. However, as their relationship deepened, she noticed a pattern of controlling behavior and emotional manipulation. John would often belittle Emily's achievements and dismiss her feelings, leaving her feeling inadequate and unloved. Despite her efforts to please him, Emily struggled to maintain her sense of self-worth in the face of John's relentless criticism.

Recognizing the Signs of Psychopathy

Understanding the signs and red flags of psychopathy is crucial for protecting oneself in relationships with potentially harmful individuals. While not all psychopaths exhibit overtly antisocial behavior, there are common traits and behaviors to watch out for.

Case Study 3: David and Lisa

Lisa met David through mutual friends and was initially drawn to his confidence and charm. However, she soon noticed a pattern of deceit and manipulation in their interactions. David would frequently lie about his past and engage in risky behaviors, leaving Lisa feeling uneasy and distrustful. Despite his attempts to charm her, Lisa remained vigilant and ultimately ended the relationship to protect herself from further harm.

Coping Strategies for Living with a Psychopath

Living with a psychopath requires a combination of self-awareness, boundary-setting, and support from others. While it may be challenging to disentangle oneself from such relationships, there are strategies that can help individuals navigate the complexities of living with a psychopath.

1. **Establish Boundaries**: Set clear boundaries and communicate them assertively with the psychopath. Maintain autonomy and protect your emotional well-being by refusing to tolerate abusive or manipulative behavior.

2. **Seek Support**: Reach out to friends, family members, or support groups for validation and guidance. Surround yourself with individuals who understand your experiences and can offer empathy and support.

3. **Practice Self-Care**: Prioritize self-care activities that promote emotional and psychological well-being, such as exercise, mindfulness, and creative outlets. Nurture your sense of self-worth and resilience in the face of adversity.

4. **Educate Yourself**: Learn more about psychopathy and its impact on relationships through books, articles, and online resources. Understanding the dynamics at play can empower you to make informed decisions and protect yourself from further harm.

5. **Consider Professional Help**: If the relationship becomes intolerable or poses a threat to your safety, consider seeking guidance from a therapist or counselor experienced in working with individuals affected by psychopathy. Therapy can provide a supportive environment for processing emotions and developing coping strategies---

Conclusion

Living with a psychopath presents unique challenges and complexities that can test one's resilience and emotional well-being. By understanding the dynamics of psychopathic relationships, recognizing the signs and red flags, and implementing effective coping strategies, individuals can navigate these challenging situations with greater clarity, self-awareness, and resilience. Remember, you are not alone, and support is available to help you navigate the complexities of living with a psychopath.

——

Chapter 15: The Future of Psychopathy Research

As our understanding of psychopathy continues to evolve, what lies ahead for research and practice? This final chapter explores emerging trends, unanswered questions, and potential directions for future research in the field of psychopathy.

---

1. Advances in Neuroimaging Technology

Neuroimaging techniques such as functional magnetic resonance imaging (fMRI) and diffusion tensor imaging (DTI) have revolutionized our ability to study the neural correlates of psychopathy. Researchers are increasingly using these tools to explore the structural and functional abnormalities in the brains of individuals with psychopathic traits. For example, a recent study using fMRI found that psychopathic individuals exhibited reduced activity in brain regions associated with empathy and moral reasoning when presented with moral dilemmas, highlighting the neural basis of their impaired moral decision-making processes.

---

2. Genetics and Epigenetics of Psychopathy

The role of genetics and epigenetics in the development of psychopathy remains a topic of active investigation. Recent genome-wide association studies (GWAS) have identified genetic variants associated with psychopathic traits, providing insights into the genetic underpinnings of the disorder. Additionally, researchers are exploring the role of epigenetic mechanisms—such as DNA methylation and histone modification—in modulating gene expression patterns linked to psychopathy. For instance, a longitudinal study found that childhood maltreatment was associated with alterations in DNA methylation patterns in genes related to stress response and emotion regulation, highlighting the potential interplay between environmental factors and epigenetic modifications in shaping psychopathic traits.

---

3. Integrating Multimodal Data

Advances in data integration and computational modeling are enabling researchers to combine multiple sources of data, including genetic, neuroimaging, and behavioral measures, to create comprehensive models of psychopathy. By integrating multimodal data, researchers can gain a more nuanced understanding of the complex interactions between genetic, neural, and environmental factors underlying psychopathic traits. For example, a recent study used a multimodal approach to identify distinct neural subtypes of psychopathy based on patterns of brain connectivity and genetic risk factors, shedding light on the heterogeneity of the disorder and potential avenues for personalized treatment approaches.

---

4. Early Identification and Prevention Strategies

Early identification and intervention are critical for mitigating the negative outcomes associated with psychopathy. Researchers are exploring novel approaches for early detection of psychopathic traits in childhood and adolescence, with the goal of implementing targeted prevention and intervention strategies. For instance, a longitudinal study found that early signs of callous-unemotional traits in preschool-aged children predicted later antisocial behavior and psychopathic traits in adolescence, underscoring the importance of early intervention programs aimed at addressing behavioral and emotional dysregulation in at-risk youth.

---

5. Translational Research and Treatment Approaches

Translating research findings into effective treatment approaches is a key priority in the field of psychopathy. Researchers are exploring innovative therapeutic interventions targeting specific neurobiological and cognitive mechanisms implicated in psychopathy. For example, a recent clinical trial investigated the efficacy of cognitive-behavioral therapy (CBT) combined with transcranial direct current stimulation (tDCS) in reducing aggressive behavior and improving emotion regulation skills in individuals with high levels of psychopathic traits. Preliminary results suggest that this combined intervention may lead to significant reductions in antisocial behavior and improvements in social functioning, highlighting the potential promise of translational research in the treatment of psychopathy.

---

6. Ethical Considerations and Public Policy Implications

As research on psychopathy advances, ethical considerations and implications for public policy become increasingly important. Questions regarding the appropriate use of genetic and neuroimaging data, the balance between individual rights and public safety, and the stigmatization of individuals with psychopathic traits require careful consideration. Researchers, policymakers, and mental health professionals must work collaboratively to ensure that advances in psychopathy research are ethically sound and socially responsible, with a focus on promoting the well-being and rights of all individuals, including those with psychopathic traits.

Conclusion

The future of psychopathy research holds immense promise for advancing our understanding of the disorder and developing more effective strategies for prevention, diagnosis, and treatment. By embracing interdisciplinary approaches, harnessing technological innovations, and fostering collaboration across scientific, clinical, and societal domains, we can work towards a future where individuals with psychopathic traits receive the support and interventions they need to lead fulfilling and pro-social lives. As we continue to unravel the complexities of psychopathy, let us remain guided by compassion, empathy, and a commitment to promoting human flourishing for all.

—

Psychopathy and Serial Killers: A Deeper Look into the Minds of Murderers

Serial killers have long captivated the public's imagination, often portrayed as mysterious and malevolent figures lurking in the shadows. At the heart of many serial killers lies the chilling presence of psychopathy—a personality disorder characterized by a lack of empathy, shallow emotions, and a propensity for violence. In this exploration, we delve into the intricate interplay between psychopathy and serial killing, examining notable examples and case studies that shed light on the dark corners of the human psyche.

---

1. Understanding Psychopathy in Serial Killers

Psychopathy serves as a fundamental component of many serial killers' psychological makeup, influencing their motives, methods, and patterns of behavior. Psychopathic individuals possess a distinctive set of traits, including manipulativeness, grandiosity, and callousness, which can fuel their homicidal tendencies and disregard for human life.

Case Study: Ted Bundy

One of the most infamous serial killers in history, Ted Bundy exemplifies the intersection of psychopathy and serial murder. Bundy's charm and intelligence masked his dark impulses, allowing him to lure unsuspecting victims to their demise. His lack of remorse and ability to compartmentalize his actions are hallmark traits of psychopathy, underscoring the profound influence of this personality disorder on his predatory behavior.

---

2. The Evolution of a Serial Killer

Serial killers often exhibit a pattern of escalating violence and deviant behavior over time, reflecting a progression fueled by psychopathic tendencies and underlying psychological pathology. As their crimes intensify, serial killers may become increasingly detached from reality, driven by a primal urge to exert control and dominance over their victims.

Case Study: Jeffrey Dahmer

Jeffrey Dahmer's descent into depravity offers a chilling glimpse into the evolution of a serial killer. Initially fueled by fantasies of control and domination, Dahmer's psychopathic nature compelled him to engage in acts of torture, murder, and necrophilia. His ability to conceal his true nature from those around him underscores the deceptive facade often maintained by psychopathic individuals, enabling them to perpetrate heinous crimes without arousing suspicion.

---

3. The Psychology of Power and Control

For many serial killers, the act of murder serves as a means of exerting power and control over their victims, satisfying their insatiable appetite for dominance and manipulation. Psychopathy amplifies this desire for control, driving serial killers to orchestrate elaborate schemes and rituals designed to prolong their sense of omnipotence.

Case Study: John Wayne Gacy

John Wayne Gacy's reign of terror epitomizes the psychology of power and control inherent in many serial killers. Gacy, a respected member of his community, used his charismatic persona to lure young men into his grasp, where he subjected them to unspeakable horrors. His ability to maintain a facade of normalcy while harboring sadistic impulses highlights the deceptive nature of psychopathy, enabling individuals like Gacy to evade detection for extended periods.

---

4. The Role of Trauma and Childhood Adversity

While psychopathy is often associated with genetic predispositions and neurobiological factors, the impact of early trauma and childhood adversity cannot be overlooked in shaping the development of serial killers. Traumatic experiences during formative years may contribute to the erosion of empathy and moral conscience, laying the groundwork for psychopathic traits to emerge and flourish.

Case Study: Aileen Wuornos

Aileen Wuornos's tumultuous upbringing underscores the profound influence of childhood trauma on the trajectory of a serial killer. Abandoned by her parents and subjected to abuse and neglect, Wuornos's early experiences instilled in her a deep-seated resentment and mistrust of others. Her descent into violence and murder reflects the devastating consequences of untreated trauma and the toxic interplay between psychopathy and environmental factors.

Conclusion: Unraveling the Enigma of Psychopathy and Serial Killers

In the shadowy realm of serial killers, psychopathy looms large as a defining feature of the most notorious offenders. From Ted Bundy to

Jeffrey Dahmer, the chilling convergence of psychopathic traits and homicidal tendencies offers a window into the darkest recesses of human nature. As we confront the complexities of psychopathy and its role in serial murder, we are reminded of the urgent need for greater understanding, prevention, and intervention to stem the tide of violence and safeguard society from those who dwell within its darkest depths.